The Startup Book

Entrepreneurship - Training Manual

By

Bilal Ahmed

@2014 Bilal Ahmed, All rights reserved.

Table of Contents

Startup CEO
Fast Interviewing
Startup Partner
Living in a World of Stories
Working Weak Ends
The Art of Concentration
Programmers Turning CTO
Entrepreneurs and Schools
Anti Aging Life
Passion
Programmer Mistakes
Startup Programmer
Quietness
Raising Worth
CEO or Startup?
Elevating Developers
Entrepreneurship Tests
Twitter Productivity
Finding Co-Founders
Computer Programmers Secrets
Entrepreneur Problems
Part-time Entrepreneur
Launching Products
Quora Power
Sean Parker & Mark Zuckerberg
Young Startup CEO
Two Careers
Quitting your Job
Venture Capitalists
Startup Mentors and Advisors
Startup Pivot: Leaner than #LeanStartup
Startup life: Direction, Execution and a pack of cigarettes
Early Stage Startup Marketing Strategy
Bootstrapping a startup
Coding a startup
myTweetmark Analytics!
Startup Pivot (Infinite loop)
Psyche of a tweet vs auto tweet
Unfollow users that don't follow back
Tweet Marketing
Startup metrics: Viral factors, Retention cohorts and Engagement metrics

- Viral Factors
- Startup Motivation
- Viral factor measurement
- Referrer Tracking for analyzing traffic
- Retention measurement
- Launching a Product
- Building Custom analytics

Introduction

Building a startup is tough. There are many battles to fight. There are many things to learn. Many of us have dealt with investors so this is a great opportunity to be humble no matter how smart we are. The time is ticking. There is much to be done. At the end, we have the ability and opportunity to transform into something completely different.

A startup is about solving problems. Invent an offering that has never been done before. A successful startup is a paradigm shift. Everyone benefits from the venture. The angel investors and venture capitalists closely monitor the progress. Building a product with the best execution is the end goal. The customers have to be communicated with, make them part of the execution model. Based on feedback, the application model pivots. Every change needs to be considered with revenue model. Who will pay for this service? Is it one time charge model, monthly subscription or pay per click? All these questions must be addressed as early as possible. What is the target audience? Are these casual users or business customers? The user acquisition model changes dramatically once we see the full picture.

Startups test our DNA. The entrepreneur faces many mental and emotional barriers. Friends become foes. There is lots of competition. Both from our loved ones that are witnessing our growth and also business startups. In the midst of everything, you have to find the best to help you. May it be coworkers, customers or investors. You also have to meet the industry greats in person. I have met founders of both twitter and facebook in person, Mark Zuckerberg and Ev Williams. There are some great minds, investors that shape the face of many greatest startups out there to learn from. Unless we make contact with them, it's hard to determine our full potential.

Startup CEO

Being a Startup CEO is fun. To build a product that is being used. It's easy to measure when the content gets distributed. Twitter and Facebook are easy products to test the marketing and distribution.

But it doesn't come easy. There is lots of work. You make your own luck. Things get difficult. You learn to be self-sufficient by understanding the value of money. There are various hats that you put on.

You develop social intelligence. How to read people and ask the right questions. Answer the toughest questions. Your network gets larger and larger. Your vision changes.

DotCom Bubble

After the crash of dot com bubble and losing about $25000 cash in Stock market, most folks latched on to jobs that would pay bills. Many visa holders were sent back to their countries. It worked out great for India and China where the same talent pool turned entrepreneurs. Now their respective industries are booming.

In a couple of years, I was lucky to find work in social media startups, pre facebook and twitter. This is the second bubble. Folks are not talking about it yet because we are still in initial phases. I was lucky to build my own startup in this sector, taking the angle of helping business with social media.

Facebook and Twitter are changing the world. The shift will be discussed for centuries to come. But for now, let's enjoy this bubble.

Fast Interviewing

I have looked at about 1000 resumes in the past 8 months, interviewed about 500 people on phone screens. From that invited about 40 people on site for a face to face interview. My phone interviews average about 12 minutes. The call is scheduled for 30 minutes. The folks that do well get extremely happy in 12 minutes. Some go on to find continued happiness by finding a better offer somewhere else. Some get pissed off because of not given enough time. But in either case, it's a win win situation. If you don't get a good response, it's time to step up. I currently manage 10+ solid developers.

Local Food Movement

We have been utilizing twitter and facebook for local food and farmers market for a couple of years. The service is free to help promote organic, fresh and healthy food options for the general public. The service also allows the local food community to do online marketing which is very expensive for them. The idea started by interviewing business in farmers markets. How much money a month they previously spent for advertising and what they received? The answer was $100 and nothing they could measure. So as a business model, a $30/month subscription fee for unlimited marketing seems like a good start for the venture to begin.

Entrepreneurship Schools

I regret not being exposed to entrepreneurship world. The place of creation is a sacred place. Only few are allowed. It takes skill, motivation and will. Action speaks louder than words. When we are going to school, only a few are lucky to stick with their major. Upon graduation, we work but none of this leads to entrepreneurship.

Now I understand better that you have to go to specific schools for that, e.g. Stanford, MIT. You also need to network with Incubator and accelerator. There are many startup events in the bay area, California. You learn more about business from network events than any book you will ever read. The network events are even more fun when you can meet venture capitalists.

Startup Partner

A Startup partner needs to be able to:
1. Focus.
2. Getting things done.
3. Understand the value of time.
4. No excuses.
5. Creative.
6. Can do; not can't do.
7. Motivated.
8. Skill.
9. Will.
10. Understand money.
11. Supportive.
12. Fun.

Living in a World of Stories

We live in a world of stories. Everyone starts with the same story. Based on how much is actually achieved is a different story. Frustration and depression occurs if we are not able to finish the story. Hence it is more important to live in the moment rather in the story. The situations we need are right in front of us. How practical are we to capitalize on those. Or are we living in the projection of another story? We have the ability to create our real story.

Entrepreneurship is about rewriting your story. But this time, it is all based on effort, passion, drive, motivation, will and creation. Focus on the problem that you are solving for your startup. Your life story will develop itself.

Working Weak Ends

In my bootstrapping days, I learned a lot from Twitter. I spent sometimes 20 hours a day, following influential people like Dave McClure on Twitter. Twitter provides an amazing way to jump barriers in your social circuits. You can learn from other groups, and over time, increase your network. I worked a lot and so did Dave McClure. One day, Dave told one of his colleagues on Twitter about "Weak ends". He says, "weekends are for weak ends".

The life of a founder is such that no matter how many hours are spent, there is still more work to be done. There are many things to address. There are many people to answer. As the whole world watches, most wanting to see failure, the pure entrepreneur keeps rising. Believing in self and the vision. Working closely with Founders is a huge blessing. I remember myself working 120 hours a week, 40 hours on a weekend, 20 hours a day. Breaking all records of working.

The Art of Concentration

Computer programming requires concentration. The more the concentration, the better the programming. Especially when all the corner cases are being thought of. There are many things to consider. The world of coding is a world of it's own. There are many issues in a multiprocessing mind. Missing one issue could mess up everything. Good programmers don't make mistakes. Hence a time is needed when the best concentration can occur. Usually that means either at night or early in the morning. When the world is not paying attention, the geek goes nuts.

Programmers Turning CTO

There is a correlation if you are a great developer, turning into a CEO or CTO of a company. When I was working at a social media company, predecessor to Facebook. I was a 5th engineer of the startup, working directly under the guidance of the CTO. One day he took me aside to a coffee shop and said, "Bilal, one good engineer equals 10". That really sunk in with me. I never looked back and coded like crazy. I moved to another startup, also created by a bootstrapping founder that made millions. Being close to accomplished founders puts a monkey on our own backs. Hungry and looking for opportunity, I started coding my own startup on nights and weekends as I saw facebook coming and eating social media startups for lunch. In the midst of everything, I created my own startup.

Entrepreneurs and Schools

Unfortunately, the opportunity of entrepreneurship is given to very few. Although it seems attractive from outside. It's actually a painful process. No pain, no gain. The rewards are huge. You get to do multiple things at once. You can't go to school for this. You have to know how to gamble.

There are schools like Stanford and MIT that have great entrepreneur programs connected directly to Venture Capital circuits. The exposure to them is invaluable. Nowadays there are many Incubators. Learning from them is priceless.

However true entrepreneurs work from within. Past all the schooling.

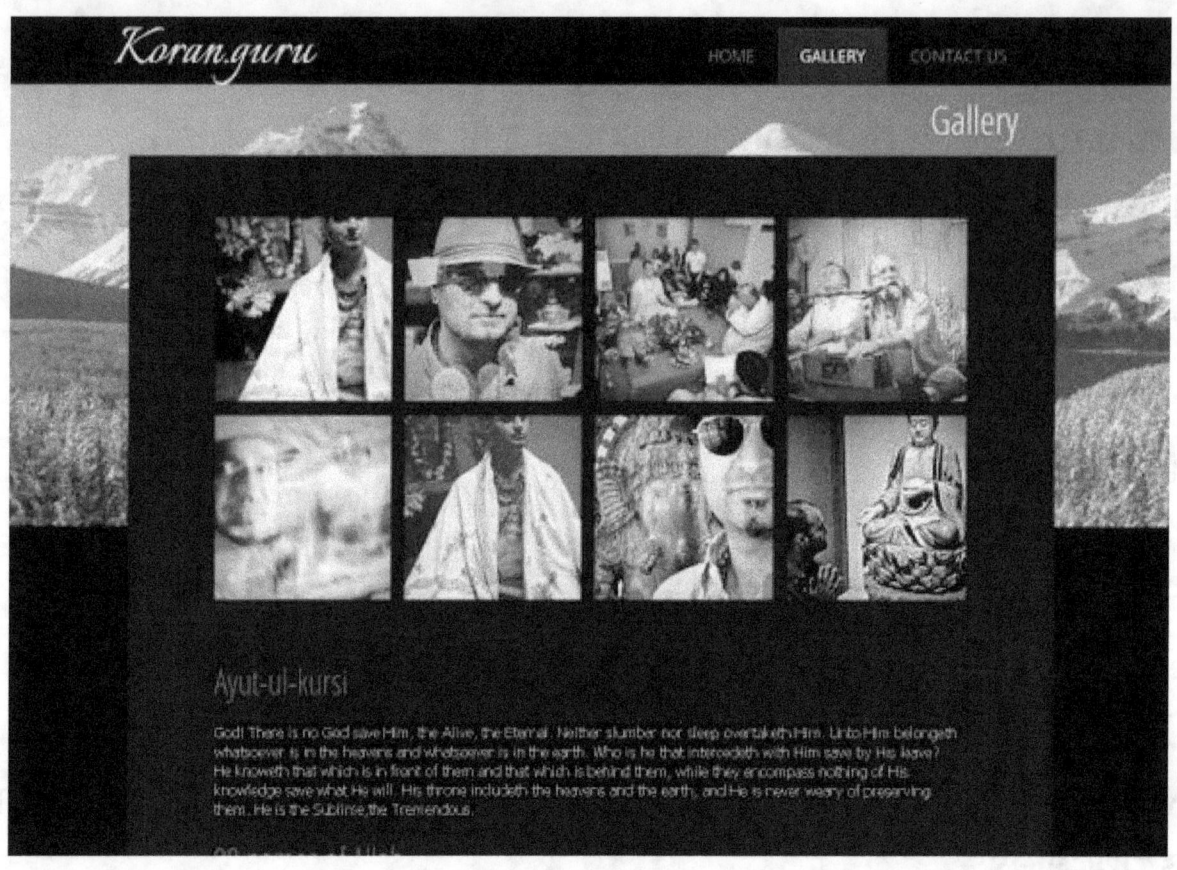

Anti Aging Life

It is never too late for anything. The key is to understand yourself and pivot to the next level up. If you get stuck with jobs and turn 40, now you are already thinking retirement. The fears "who is going to hire me?". The options become limited. The worst situations show up as ugly faces. Knowing you are down and they will push you down further.

Rather, take on a different spin. Learn. Take care of your body. Don't sit around, rather create opportunity. Be hungry. Stop complaining. Build passion for things you enjoy that can generate income. Be happy.

Passion

Working at a job teaches us skill. However more often than not, once we elevate in our career, we realize that we are just a little bit in the grand scheme of things. Then we start talking ideas but having very little skill for any real manifestation. So therefore passion can't fully bear fruits. The real measure of passion is entrepreneurship. People jump job to job, never fully established in their skills or careers. The best is to go solo and work on own ideas. Build passion. Then the work you do, whether your own or for someone else will be passionate as well.

Passion is a place of effort and love. Keep increasing your passion, and you will look for things to create. There is amazing feeling that happens when you are able to see your idea in manifestation. That creates the passion to build bigger and better things.

Programmer Mistakes

The biggest mistakes could be many. While the programmer is busy head down solving problems, the world creates all sorts of judgements towards the character. For example, "she doesn't talk much". Really, that's a problem? While she is busy thinking about real world problems? The programmer as a result shuns down, looking for confidence and internally figuring out a way to prove self.

The way to fix this is do side projects at home. Become an entrepreneur, learn business and overcome weaknesses.

Startup Programmer

If you are a programmer and realize that you can become an entrepreneur, the advantages are huge. Once a developer gets exposed to entrepreneurship, they can see patterns no one else can. Because the programmer can take the problem in the logical sphere. The business entrepreneurs look for patterns of opportunity by looking at the market. A programmer tends to solve business problems by going inward. Eventually the market opening matches the inside intuition which was the case with Mark Zuckerberg, the Facebook founder.

Quietness

Programmers are introvert. We live in our own world of logical patterns. Everything to us must have a reason. We are always solving problems. Solving a puzzle is more important than communication on conversations. The experience is very close to a yogi that just lives in own realization. Programmers can become great teachers of meditation.

Business people are extroverts because they have to interact with partners and customers. The best combination is found when synergy is created between two or more partners. It is very important to understand each other's strengths and weaknesses. Stay quiet and be more focused.

Raising Worth

As a programmer, there are milestones for a raise.

1. Can you write code?
2. Can you debug?
3. Can you isolate the issue?
4. Can you reproduce the problem?
5. Can you instrument the code?
6. Can you take the design specification and implement a feature?
7. Can you work with a team?
8. Can you lead a team?
9. Do you work on outside work projects?
10. Are you an entrepreneur?

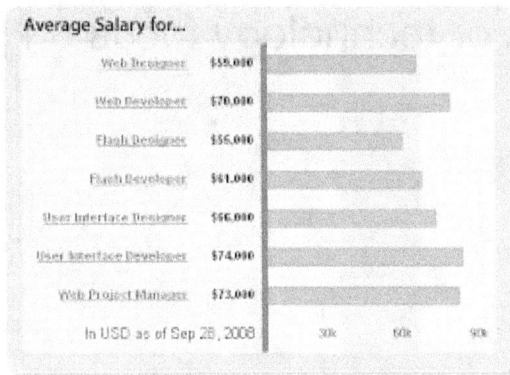

CEO or Startup?

If you can't code in a startup, what can you do? Generate lots of sales? Attract whole bunch of investors? Good at paper work? You still have to build a skill that generates value.

If the above paragraph seems pessimistic, that was the goal. So many people focus on title and completely miss the mark. Startups are about doing. There is no point in showing off yourself. Show off your product.

It's important to have charismatic founders that can pitch in front of investors. However it's more important to always be brilliant at what you do. Hence successful founders, whether business or technology focus on product for their startup.

Elevating Developers

You must question your motivation if you are a good programmer. Are you doing work for money? That gets boring. Are you doing for title? That's all good till your peer gets ahead. Are you doing it for passion? Aa haa! Now we are talking. Now you could be the next Steve Jobs. If you believe in yourself. Good luck!

I have a lot of respect for developers that have the passion to become entrepreneurs. It is an amazing feeling where the art of development is mixed with building something important to the world. The place where the heart meets the mind.

Entrepreneurship Tests

No external motivation is a moment that tests our ability immensely as an entrepreneur. You learn a lot from it. When you are in the process of climbing the ladder, friends become foes. The detachment from old associations is a tough path. However as we realize the differences and carve the new path. The results are much clearer. You learn to believe in the vision and work towards momentum. The momentum is what leads to a place much more satisfying and sure about yourself and your decisions. The real answer is within.

Twitter Productivity

Twitter in our age is one of the most productive way to spend time on the Internet. Initially, I spent a lot of time on twitter. There are walls broken in terms of getting to the people needed to help learn our next stop over. You can tweet to celebrities. They might not respond but they will think about you. There is an energy exchange that takes place. I have gotten response back as well which is always a pleasurable validation. I followed some key startup investors and market movers and build a product which I thought they would like. I can true to its validation.

Finding Co-Founders

You don't just find co-founders. You have to network in to see if one is available. The next part is synergy. The vision has to be aligned. Skills have to be matched. You have to establish divide and conquer to not step on toes. Awareness is developed. Respect needs to be established. Love, not fear is needed to allow creativity to flow. As a technical founder, I have trouble finding business co-founder that can match my skills hence I chose to be a single founder.

Computer Programmers Secrets

Computer programmers know where the problem is. We are trained to look for them. The question is how we react to them? A lot of times programmers get in the cross fire with business for that reason. Business is about results. Bad programmers get stuck in conflict. The key is to display results.

Entrepreneur Problems

Entrepreneurship is all about solving problems. When we reach one height, there is always another hill to climb.

Growth doesn't come easy. You have to be fearless in its pursuit. Swami Vishnudevananda came to San Francisco from India to spread yoga. Fifty years later 35000 yoga teachers paying thirty five hundred today per teacher training.

Part-time Entrepreneur

It is possible to be a part-time entrepreneur. It's not about time. It's what you do with it. The standards in the investment sector are high. Money is short. You have to prove revenue or show a hockey stick. Find a day job or only work as an entrepreneur if you are getting paid.

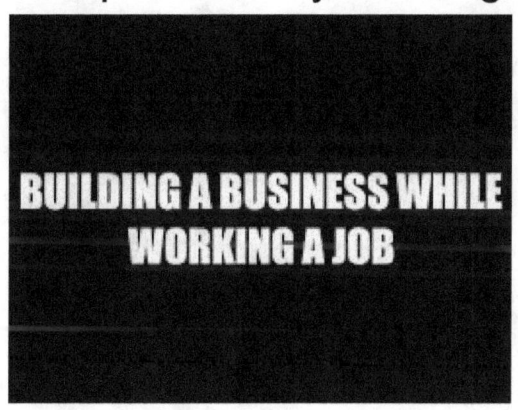

Launching Products

Never be shy of launching. If you are working to please others, you will have blocks. If you work to please your work, there will be rewards. If you surrender your work for greater good, there will be great surprises. Bigger rewards. So just focus on your vision and create. Ignore the distractions.

Quora Power

Quora is a powerful tool to help you convey your message to a wider audience and measure the influence. The communication with the community helps define whether the subject matter to discuss has value.

The micro blogging on quora can help you get started with aggregating content. Once you have enough content, you become a subject matter expert in your field. This work can eventually turn into publishing books.

Sean Parker & Mark Zuckerberg

Sean Parker was brilliant. He helped structure Facebook relative to investors & shareholders in a way that ensured Zuckerberg was always in control of Facebook. Understanding human psychology to that depth doesn't come easy. Given the background of Parker from the fall of napster. Trusting Mark and the vision of Facebook. After helping facebook, Sean continued to do a couple of other social networks as well. None close to Facebook success but still they did fairly well in exits.

Young Startup CEO

The best advice for a young, first-time startup CEO is to be hungry. When the hunger comes from within, it creates a powerful determination from within that is unstoppable. The ability to be firm and correct. The ability to create follows. All the doubters become less important. You know from within that once you satisfy your hunger, you can feed many from it.

Two Careers

Of course, you can pursue two careers at the same time. The day job is important to bring in money unless you are VC funded. Then your job becomes your passion. For bootstrap startup, the earning is needed to fund your startup. Bootstrapping a startup is much more fun because the rules are much broader to what you work on. This is where the creativity flows. The VC backed startups can feel like a job if too micro managed. The most important factor is always about what a job can do for us to build three main ingredients for success: motivation, skill and will. This builds the confidence for "anything is possible".

Quitting your Job

Rather focus on fear, focus on opportunity of quitting your job to start a startup. Once you have already established that your path is of an entrepreneur, then focus on what are the key ingredients of starting a business.

A job only provides short term happiness. You might start off great. Everybody gives you a helping hand at first. However as time moves on, and if you are an entrepreneur caliber person, chances are that you start reaching heights in the corporate world very fast. Once that happens, you see competition which is detrimental for your growth. Meanwhile, cash is coming in part of your salary.

The short term pleasure is great when the money comes in, however there is a direct correlation to your health.

Then the opportunity of entrepreneurship arises. You see the excitement and pleasure it brings, much more potent than any enjoyment you ever felt. The joy of creating a product that is being used by others is overwhelming.

Venture Capitalists

You can't learn in school what most venture capitalists know. You can't learn in school what entrepreneurs with passion, depth and knowledge know. The knowledge comes from deep grind and a can do attitude, rather can't do. To be the best or give someone else a chance. The will to do whatever it takes to win for a larger goal and bigger prize.

Finding the right investors is critical for the success of the startup. The network of investors can generate any amount of money for the well being of the company. They can unblock pathway by rigorous research and understanding competition. Once a high caliber startup is founded, due to the fact that the many creditable venture capitalists have invested in it, it's an easy target for exit.

Startup Mentors and Advisors

Startup life is hard. The road is long and there are many roadblocks. Mentors and advisors are important aspect of the journey. Below I will outline keys to finding successful mentorship:

Available: Startup progress is a game of timing and momentum. How available an advisor is makes a huge difference whether they are able to help you at critical junctures.

Synergy: Do you drink beer with your advisor? Not literally, but are you on the same page? Do you look straight in to eyes of your advisor?

Previous experience: What is the qualification of the advisor? Do they have resume credentials to help you unblock the road.

Domain expertise: How much do they know in the area of your business. Can they help you on technical side or the business side?

Outdated Skills: At a recent visit with Tim Draper, the founder of DFJ, he pointed out Mohr's law where in 5 years, we will make a leap in technology and business of what we achieved in 50 years in the past. Is the advisor still part of new wave of startups or their skills outdated?

Early stage startups require a completely different skill than late stage startups.

Motivation: What's the motivation behind the advisor helping you? Is it corporate titles, money or passion. Passion lasts longer than everything else.

Introductions: How eager are they to make introductions to people that can unblock the road and help you grow your business?

Big picture: Do they see the big picture and your business growth or are they just hanging for the ride?

So there you have it. Choose your advisors and mentors properly and put them on your weakness. If all the above are answered correctly, you will enjoy the ride and I am sure they will be part of your success as well.

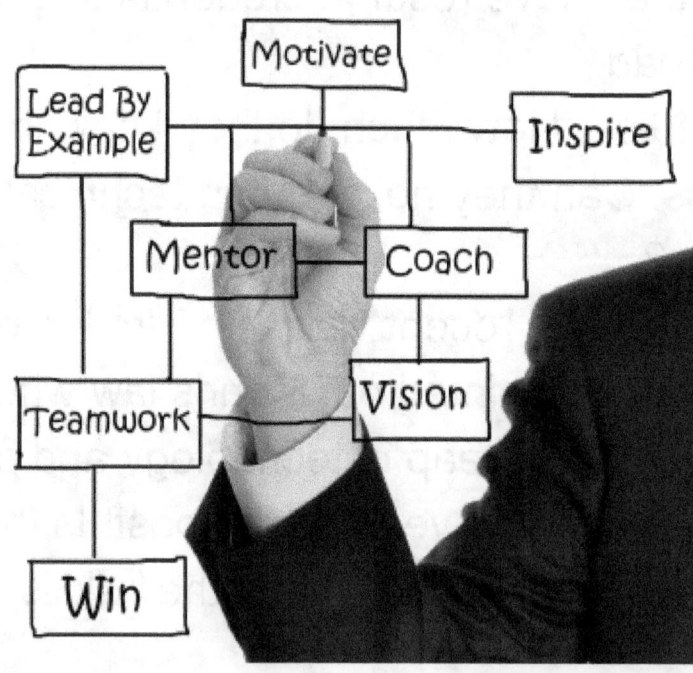

Startup Pivot: Leaner than #LeanStartup

My company, @mytweetmark has launched many products and websites. Fail fast, pivot and try again. Some of them died, some of them are doing great but more coming! Here is the @mytweetmark lifestream:

Highlights:
January 2010: @mytweetmark launched - Saw need for blogging and metrics, cheaper advertising option to Google and Facebook utilizing open Twitter API.
January 2010 - March 2010: Cloned @mytweetmark to 5 sites:
@mytweetmark - Auto tweets
@drspiritcom - Spiritual thought sharing site
@mytweetsports - Sports score sharing site
@viralfactors - Viral analytics
@homecookme - Recipe sharing
March - April 2010: SXSW affect, killed @drspiritcom, @mytweetsports and @viralfactors! Chirp effect, chipmunks coding away at chirp. Many implementing similar ideas to @mytweetmark..
June - March 2011: First mover advantage in signing up all the local businesses for over 150 farmers' markets in Northern California and Austin. Not signed are still on email campaigns, receiving offers.

March 2011: SXSW affect, launching @winehomeme for wine sharing and iphone mobile applications!
August 2011: @homecookme mobile
August 2012: Relaunched @drspiritcom
April 2013: @TheSecretBookOM published
June 2013: @TheSacredBookOm published
March 2014: Relaunched @mytweetmark and @homecookme
May 2014: Reintroduced blogging platform, to go with auto tweets.
July - August 2014: The Startup Book

Some of the lessons learned:

Speed is everything.
Execute fast and learn on the metrics.
Make decisions.
Implementing new technology is much easier and faster.
Follow LeanStartup
@twitter is amazing marketing platform.
Never be ashamed of fail.
Quit whining, continue doing.
Action speaks larger than words.
Meet as many angels and VCs you can.
Pivot on feedback.
Time to market is everything.

Understand competition.

If your execution sucks and late, the market will be taken.

If your execution is early, determine whether the opportunity is worth the wait?

Only the first one gets the dinosaurs, i.e. the hockey stick chart or biggest piece of the pie, market share.

Startup is not about weekends, or weak ends (VC terminology). It feels like a weekend every day!

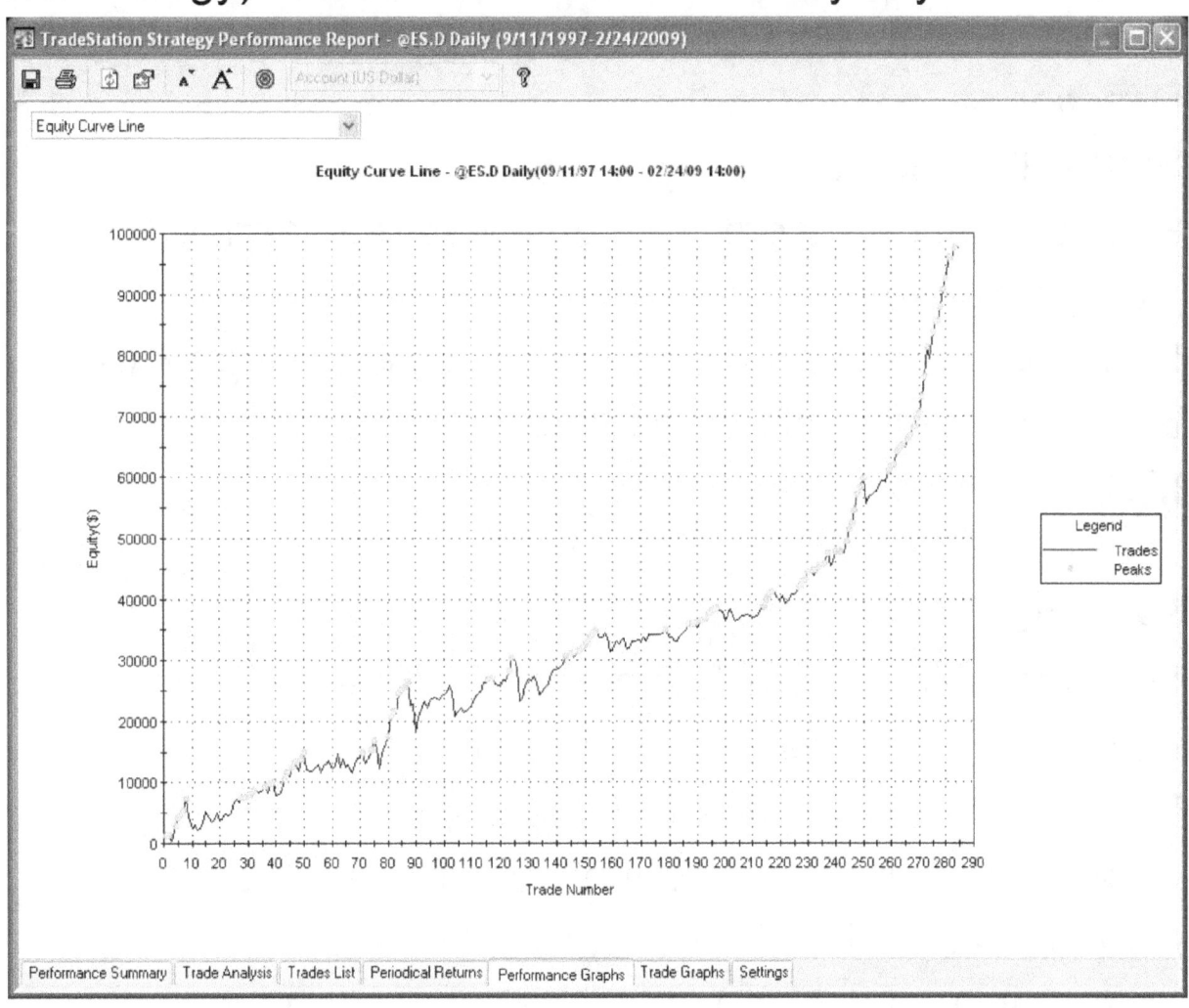

Startup life: Direction, Execution and a pack of cigarettes

Startup life is tough: You need a thorough commitment and focus. Recently I was watching NBA basketball and Jeff Van Gundy mentions inspiration of Kobe Bryant: Motivation, Skill and Will. My variation is a little different; Direction, Execution and a Pack of Cigarettes:

Direction:
Start with an idea
Pivot or change according to business requirement
Generate revenue
Profitability is achieved when business is paying for itself
Exit or merge into bigger offering
IPO

Execution:
Do it yourself because technology is easier to implement
Find folks with synergy to work with
Find amazing advisors and investors
Facebook engineer right out of college makes $120k/year in bay area, California. Architects $180k/year
Lots of attitude to hire any technical talent and the positions are in high demand
Contractors make $150/hour and more

Pack of Cigarettes:
Not literally
Skilled folks like to be comfortable
Move in herds
Do what makes you happy
Skilled people have lots of choices

Early Stage Startup Marketing Strategy

Startup marketing strategies need to be crisp, with lots of options. Here we will discuss how to implement various marketing techniques for early stage growth of your startup.

Distribution: If you have a subscription model or monetizing your audience where every conversion will result in a sale, then advertising is a viable option. If your product is not ready and you are advertising, you have a short window to start converting into sales otherwise it will kill your monthly burn rate. Once you start advertising, your audience will expect the site traffic to always be on that level and up.

Social networks: Leverage affective usage of Twitter, Facebook, Linkedin, etc. Affective social networks marketing reduces your cost of advertising.

Search: All data produced from your website needs to effectively be searchable by Google and show up high ranked and on first page of result.

Data: Capture all the data. Collect all the information you can about your users. Overtime, the data will make more sense and growth patterns can be obtained. Data helps you build more product features over time that are viable to users.

Email campaigns: More now than ever, email campaigns are huge! Email campaigns help you define your audience. Tools like mailchimp, constant contact, etc. will help you determine what users are interested to hear about your service, or not. Regular emails create a sense of community, a niche. Measure conversions, clicks, open rates will give you a great ideas of the value of your service. Are people actually excited to know about your business?

Other sites: Partner sites are key. We are moving into an arena from portal websites to niche websites. It's important to create a cluster of companies that help you build a bigger offering. Each company specializes in key areas.

Word of mouth: Word of mouth and gorilla marketing is back in action. Companies like yelp started with local buzz in local niche markets and by word of mouth and social connection, expanded to huge businesses. Connect with your customers and add them to your email campaigns mailing list to promote constant word of mouth marketing.

Measure: Measure is the buzzword for 2010. Unfortunately, most people know how to measure but don't know how to act. Once you get your analytics report, identify patterns. Is your offering too large to know about your customers? Viral factors, retention cohorts and engagement metrics. Figure it out.

So there you have it. Enjoy and happy marketing for your early stage startup!

Bootstrapping a startup

Startups are tough! Here is a blueprint that I followed building @mytweetmark and @homecookme. Your strength and endurance through the process is key. People break with pressures. However I always believe that you go after the knowledge and gain. Win or lose, nobody can take away your knowledge. Live each day like it's your last. Make each hour count towards productivity of your company. Bootstrapping is very difficult because money is small. You have to make it happen. Your self-discipline matters. Yoga, exercise, etc. helps. You have to keep producing, while having a good internal balance. Break the time into the building the product, or interacting with customers. Everything is moving at a much faster pace as the product evolves, but so are you. Just know that you are making yourself stronger day by day by going through the process. Without pain, there is no gain, right? Welcome to entrepreneurship!

Idea: Start out with a simple idea. What is needed? How can I make the world a better place? Do research. Come up with first wire frame, or screenshot of the product.

Money: Burn rate. Everybody has expenses. Nights and weekends to start out the product. Spreadsheet to calculate month by month expenses into the product.

Project Plan: Helps a lot! You can start seeing a picture of tasks that needs to be accomplished. Really helps with head down focusing and breaking things down in design. Write down the task, duration, priority, description, date, etc.

Market: What's the market size? @mytweetmark heavily utilizes @twitter therefore it was key to spend a lot of time on twitter and understand the culture, users, product and figure out how to offer additional services. I spent more time on market that anything else in the beginning. Learn about investors through Twitter because they help you understand demand and competition. Twitter has millions of users and Facebook has over a billion users. Search tags # help penetrate the market through tweets.

Validation: By building a limited feature product and deploying in front of twitter audience. Conferences help a lot, etc. #chirp, #sxsw, #f8. Also, by asking the correct questions to VC and Angel investors. Remember @drspiritcom, @mytweetsports, @viralfactors? Yup, I deployed 5 websites, and dropped 3, kept only

@mytweetmark and @homecookme. Fail fast and accept the grief quickly.

Focus: Reiterating back into the product. Found a new niche with farmers markets customers desperately needing social media. Introduced v2 of @mytweetmark. The demand utilizes both @mytweetmark and @homecookme products so no work lost.

Pivot: Changing the product so it fits the demand of the farmers markets audience.

Market: Understanding the market size, competition, execution time, etc. Farmers market is $1 billion business in United States. 16% growth in past 1 year due to economic crises. Produce brings cash. Farming community is way behind on social media. Willingness and ability to hand hold, and help the audience.

Customer Validation: Early customers using the product and giving feedback. Their like, dislike, eye popping, eye brows up, etc. signals are key. Measuring happiness.

Influence: How is your product changing the market? Can you measure the influence? Klout score helps. Also, how is your customer influencing the market? For example, TV commercial ads play over and over again. When you are ready the buy a certain product, obviously seeing the right advertising helps make that decisions. Auto tweets help us emulate that behavior for farmers and customers on Twitter. Our customers on radio station, interviewed by bloggers, New York times, etc.

Ranking: We are the largest network of social media farmers market businesses in Northern California.

Pivot: Never stop changing.

Growth: Backlog of features to implement next. How can I influence the market and grow? Klout scores to individual farmers, deploying on one forty platform, viral, iphone application, coupons, etc. are all ideas for growth coming next.

Coding a startup

Starting a company from the ground up is a challenge - Here I am going to outline the decisions and trade offs I made. Easy blue print to start coding a startup.

Getting up to speed with technology: Unless you graduated a few years ago, the curriculum and paradigms are outdated. Business students are learning about Steve Jobs from 80s or Computer Science student studying C or Java is not really practical in the work field. Pick up a few books and thoroughly get updated with technology. I picked Grails because I can always go back to Java if needed. Grails is a full on web framework with GSP like JSP, MVC, spring, hibernate, syntax mixture of Rails, Java, PHP, etc. It has built in application server, dependency injection, logging, many plugins. It also has plugins for twitter, facebook, paypal, etc. Code can be written 1/10th of lines compared to before.

Selecting hosting: I started with 1and1 hosting, but Rackspace is awesome. Great service, new company, easier product to use, cloud computing rocks and they also provide file transfer service through their cloud as well. You can upgrade or downgrade memory and CPU environment without having to reinstall anything.

Source control: Subversion SVN works great for startups. Very low overhead, lots of documentation on the internet. Many people in the industry are using it so easy to get help.

Database: I choose postgres over mypsql because I have previous experience using it. Either one is fine.

Docs: Sharing documents among team members or even contractors is never been easier than using Google docs.

Scripting: I am a strong believer in writing strong shell scripts to minimize the amount of typing and making errors. A strong script when written well never fails. I usually just login to server and type 'reinstall' to get the latest deploy from my laptop onto the server. Keep it really simple.

Process: Using continuous deployment patterns, It takes 3 minutes from the point I checkin my code into repository to code being deployment on production server. There are many shell scripts to run commands and do the validations. The key is to make the process so simple that the focus is 100% on the product development.

UI/UX Skills: Technology is getting so simple that more and more startups are leaning towards Lean Startup to make software development faster. Startups engineers are not only expected to know backend, but really understand User Interfacing design and experience. Hence html, javascript, css skills have to be sharp.

Platform: Twitter and Facebook provides great details on users using their authentication schemes. Signups are dead. People don't trust them as much. Twitter and Facebook provide really easy API to use. I use Twitter4J for all Twitter api calls because it's written in Java.

Security: Web is very insecure, hence it's important to code with security in mind. Input forms should always check for hacker code in Javascript, server should have ssh security with ports, etc.

myTweetmark Analytics!

People know I have spent a lot of time with metrics and dashboards. If I see any piece of important data, I build a dashboard around it so with charts and graphs, I can identify patterns. These patterns are usually the secret sauce for understanding your business. Yes, I have gotten people fired before for exposing too much data. But that's all because it's in the best interest of the company. Nothing personal, just business.

on myTweetmark, we are exploring the Twitter ecosystem on hashtags and reach. I could create simple categories for my content, and get instant page views because of using proper hash tags. But the fun doesn't stop there. Once the users on twitter clicks on mytweetmark link and come back to the site, we have all sorts of interesting referral data to play with. Currently, I are displaying the hosts % break down by google visualization api charts. I could however get information about country, page views, uniques, time spent, etc. from this powerful referral data. There is lots to be explored in this area. Think about it. You are driving a marketing campaign, and you want to know everything about your traffic. This allows you to fine tune your campaigns and get the most for your buck. Trust me, I spent a lot of time with marketing initiatives from engineering angles for a couple of extra large social networks. The more the data, the better the metrics.

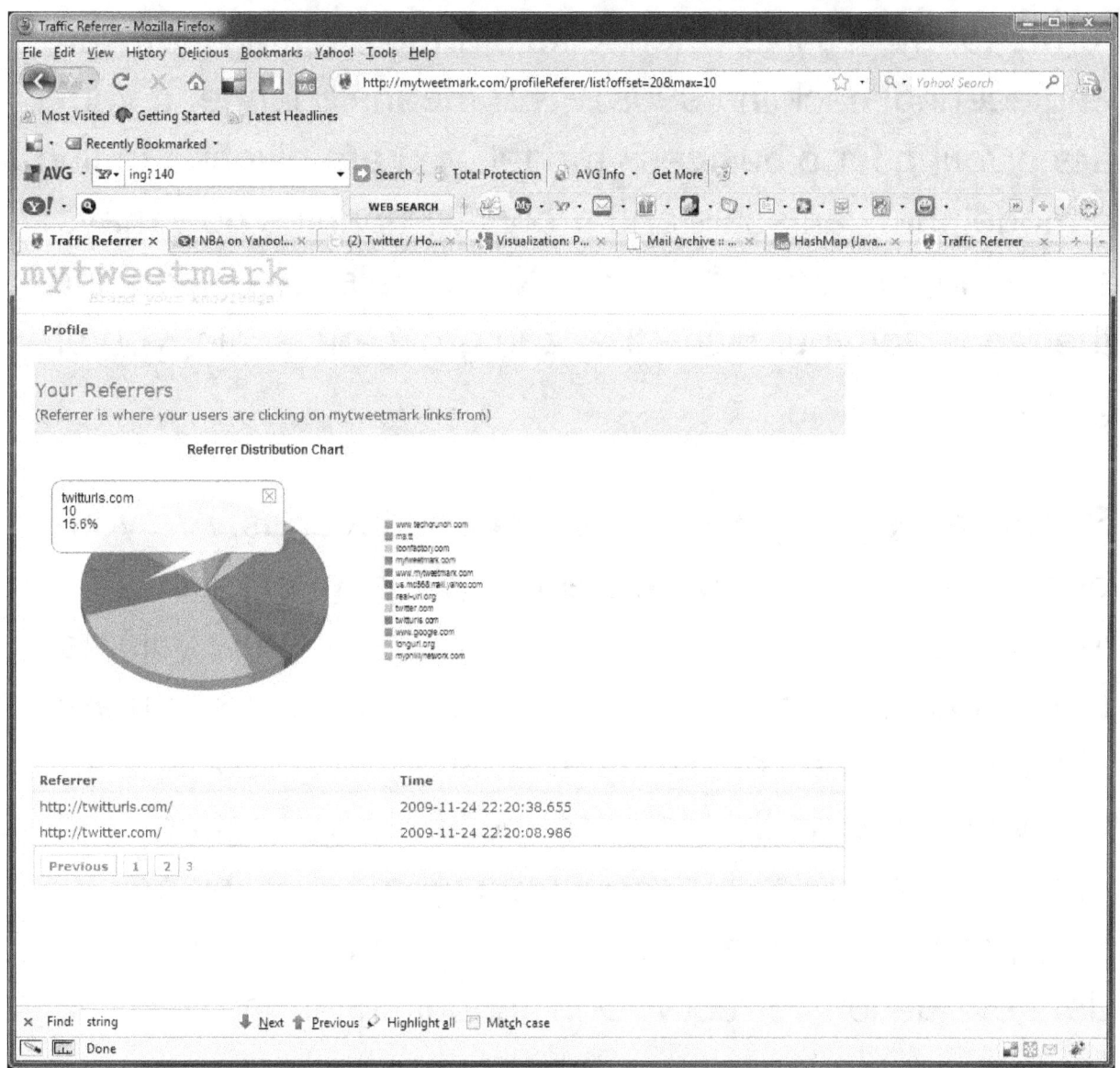

Startup Pivot (Infinite loop)

Unfortunately, 'startup pivot' is one of the most misused terms today. It's like when folks started defining 'software engineering' back in 1990s. Pivot means change. If you are pivoting for a business model, you are pivoting to find a model that can scale and grow. Companies nowadays are being built around niches. Not portals. Once you find the model for your niche that can grow and scale, you are done.

Should my employees be pivoting? It depends. Are you pivoting for design? What are you pivoting for? It's important to have clear distinction for pivot and evaluate the results. Metrics are important to install so as you are pivoting, you are constantly evaluating your data and really understand the data, not build it (engineering problem) or look at it (project management problem). Everyone in the team should understand the process of data evaluation correctly. Don't shoot shit in the dark forever. Infinite loop.

Times are changing. A new decade has begun. It's the decade of multi-skill talent people. For example, you have to know design, but also be able to develop it. If you are an engineer building metrics, you are also responsible for understanding it and figuring out how it impacts your business. Don't just shoot shit in the dark, measure the results and then make smart decisions. Understand that pivot is a great thing. It will help you resolve your problems and remove obstacles. Persistence is good and also timely execution of tests.

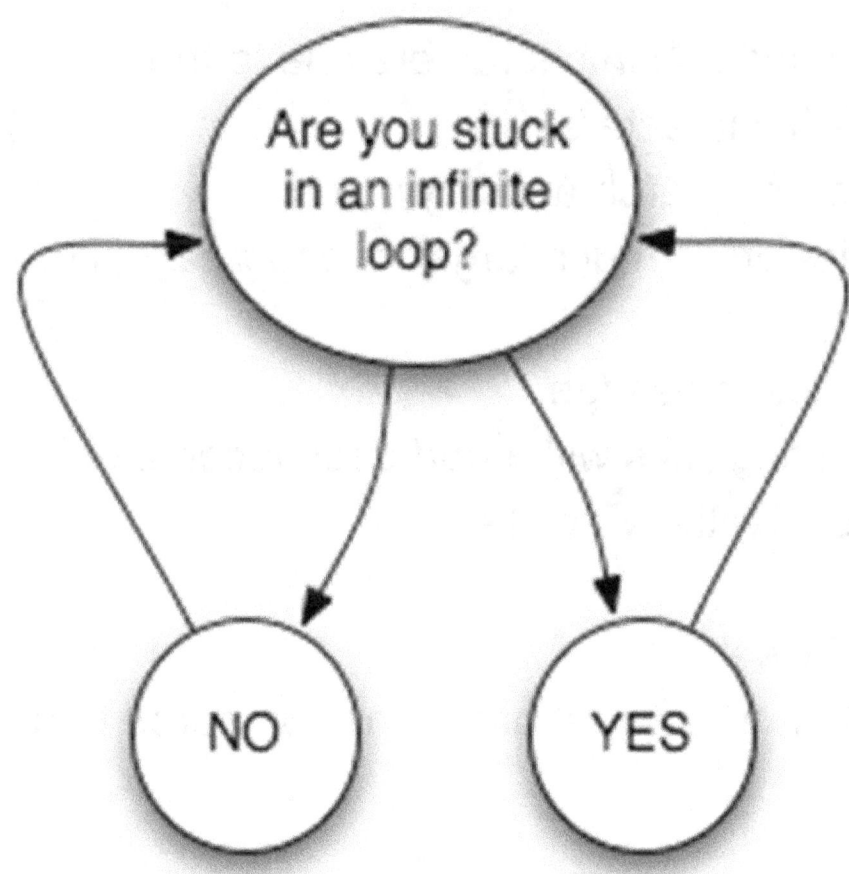

Psyche of a tweet vs auto tweet

Tweet vs Status:

The benefits of a tweet are many compared to a Facebook status:

A tweet is small and rich in characters due to the limitation, i.e. 140 characters.
It's textual, great for search and keywords.
It's radius for viewers is much larger due to # tags, i.e. conferences.
Great for marketing message.
Great for distributing links with short descriptions.
More benefits on Twitter website.

Tweet vs Auto Tweet:
The benefits of an auto tweet are many compared to a normal tweet:

Automatic, hence physical person doesn't need to be involved.
Great for repeat marketing message.
Great for distributing links with short descriptions, continuously.

The tweet stays recent, hence high score on Google search results.
Gain continuous stream of followers and views.

Technology is changing very fast. Recently at SXSW 2011, Tim Draper of DFJ mentioned Mohr's law in effect where in 5 years we will evolve 50 years growth. Lots to learn in Social media management systems (SMMS) space, however auto tweets is a great advancement from simple tweets. If you can educate your audience for it. Sometimes people don't like automatic, but that's their ignorance and preference rather any real reason. Opinions are many, we have lots of analytic data to show results and many customers recommendations and thank you letters for rewards for using auto tweets from local food and farmers' markets all over Northern California.

Most awareness happens at the beginning when the content is published. However not everyone is available to view the content, especially on Twitter. The interest goes for a while and then new content is posted. The repeat of the content posting still shows audience interest, people that didn't see the post first time through.

Communication Threshold

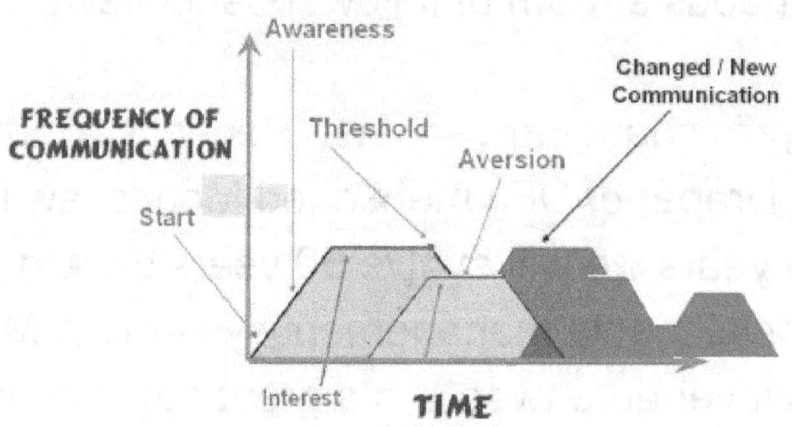

Unfollow users that don't follow back

Twitter is transforming individuals into personal brands. With your profile, users can determine who you follow, how many followers you have, how many people are re-tweeting your content, the nature of your expertise and links to your business. The users can upload images, links to videos, documents and other multimedia content to further engage their followers. We are living in an Information Age where the transfer of knowledge happens, further increasing our consciousness. We have the ability to follow people that helps us build our intelligence.

When people on twitter look at your profile, its important to send the right message, encourage them to follow you and absorb your content. The hash tags are key because it shows your users what content to expect. The auto tweets is a great way to continue engaging users, increase recency of content and allow google to index your content. Content on twitter gives high SEO score on google indexing.

To increase followers, a proven strategy is to mass follow users, based on search tags. When a user sees that you are following them, it gives them a chance to review your profile. If the search and the content type is relevant to the user, they will follow back. The conversion rate for following users with relevance is around 25%. This means there is 1 out of 4 chance that the users will follow you back. Twitter is strict about aggressive following so its important to follow only a few hundred followers at a time, for example 300. You should wait at least a week, giving the people you are following a chance to follow you back. After this time period is up, measure how many people followed back. Connect with them to encourage communication. You then need an ability to unfollow users that don't follow you back.

Twitter is very strict about unfollowing users aggressively. You should only unfollow users, based on the counts available on various twitter tools. After unfollowing the users that are not interested in your brand, wait at least a week before again searching for new users and mass follow them.

Tweet Marketing

Marketing your blog or company on Twitter is very fun! You can do all kinds of fun stuff, talk to interesting people and share content. You can make great connections and friends. You can reach so many more people by following or search tags for conferences and meetups. Twitter users are very engaging. They love constant flow of information. Here is how @mytweetmark can help you find interesting people. Please find interesting tweets that are marketing tweets. Use our auto tweets. Let's measure the impact through our analytics.

Login to @mytweetmark with your @twitter account. To test the value of your tweet by page views, go to Analytics, type in the twitter status with a link back to your @mytweetmark profile. For example, "Please try out http://www.mytweetmark.com/ for #newtwitter #analytics ". To find the best hashtag to use, please visit http://www.wefollow.com or http://search.twitter.com . Once you have measured marketing tweets, please go to Auto Tweet section and fill in 100 tweets. In the tweets, please also have links to your blog that is measurable through analytics. Please give it a month. Let's measure the quality and quantity of your followers and page views.

Startup metrics: Viral factors, Retention cohorts and Engagement metrics

Viral Factors:

A simple example is importing address book. If I import address book of 100 users and send out invites for 10000 users, the child registration conversation has to be over 100%, e.g. viral factor of 1.00 for a startup to be in a growth mode.

Build a dashboard.
Build summary tables.
Measure # of users.
invites send out.
open.
clicked.
converted.
of child registrations.
calculate viral factor: total child registered/total parent registered.
Google chart intervals of viral factors, 1 day, 3 days, 7 days, 15 days, 30 days, 60 days, 90 days.

Retention Cohorts:

Measure how many times your users are coming back to the site.
Build a dashboard.
Build summary tables.
Measure # of users (newly registered).
# of user logins.
calculate retention coefficient (cohort):
amount of logged in times/total users registered Google chart intervals of retention cohort, 1 day, 3 days, 7 days, 15 days, 30 days, 60 days, 90 days.

Engagement Metrics:

Measure by product/feature, how many uploads, viewed, etc. metrics on how the adoption is increasing or decreasing.
Build a dashboard.
Build summary tables.
Measure # of users (newly registered).
# of uploads, # of views, # of comments, etc. aggregate.
calculate engagement coefficient (cohort):
amount of users engaged in content/total registered.
Google chart intervals of retention cohort, 1 day, 3 days, 7 days, 15 days, 30 days, 60 days, 90 days.

Lessons learned:

Don't over complicate the problem: Hadoop is a solution; Metrics is about finding the problem. Learn how to read metrics and make decisions that impact the metrics directly.
Build vs. Buy: Build in-house analytics solutions. It takes only a couple of days to build internal metrics. Building metrics helps understanding of analytics and making educated decisions. Analytics companies: Google analytics, Kissmetrics, Kontagent, Mixpanel, etc. Extra points: Add country, email domain, age, gender, locale, etc. to viral factors, retention cohorts and engagement metrics to really understand your user demographic. Syslogs can be essential data to determine additional engagement metrics.

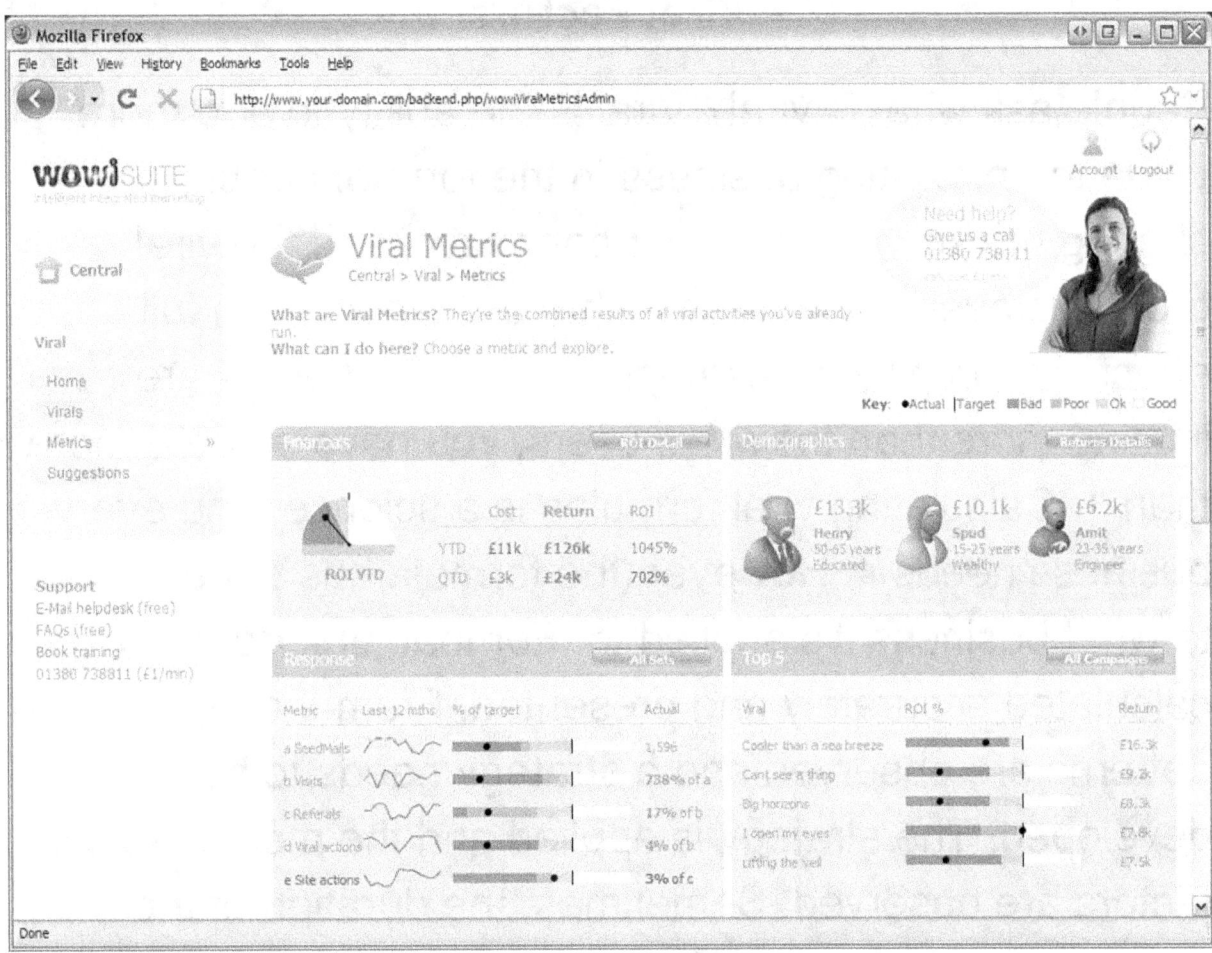

Viral Factors

People ask what is viral? Viral is the ability to add more value to an existing business in the form of people sharing the service with their connections. If the viral growth is over 1, that means that the growth is in full effect. To simplify for example, if a set of # users are inviting more than their # of users, you have a score more than 1. Calculating viral formulas is a science, and every business needs it. However, the formula has to be tailored to what business it's applied to. For that, first data is calculated accurately and presented. From these, patterns are observed and a strategy needs to be developed. The strategy is applied and the measured factors are observed. Sometimes, the duration of the factor is very important. For example, if you are sending out invites, you need to wait a couple of days for the users to receive those invites before taking their statistics seriously.

Retention is also key, because people focus so much on viral and end up spamming the crap out of their users. This creates a disjointed effect in the space, and people feel that they receive too many emails for no reason. Effective email strategy is key. Timely emails can not only increase retention, but also viral growth. Retention can simply be measured by looking at a cohort, i.e. a group of users and determine how many times they are coming back to the site. Retention can be measured not just by users, but also content, for example how much content is being published per month.

Finally, it's key to have a good grasp of your engagement metrics. This is the ability to drill down in various measurement categories, and develop a combined score. For example, for a social networking site, the amount of users, photos, friends, content, time spent, etc. are all key metrics that need to be paid special attention to. If a user wants to upload a photo album and the service fails, chances are that they will upload the photos to a different site and in the future, continue to go to that other site.

To run a successful company, all the above areas should be monitored very carefully and historic data should be looked at for trends. I highly recommend the use of tracking tables and then build summary tables. Groovy on grails is excellent for Java developers, that want to quickly look at database tables and build dashboards. Google charts is also affective way to display charts, without spending a tremendous amount of time learning charting tool kits.

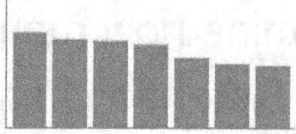

Startup Motivation

"Most men die at 27, we just bury them at 72." - Mark Twain

A Startup motivation can be measured in 3 things:

Money: Need a big pay check? Middle stage & late stage startups have bigger payouts than early stage concept startups. However, the desire lasts for about 1-2 years.

Title: If you are looking for a solid title and experience, early stage startups are much generous. Get a nice acceleration boost to your career. Large companies love startup experience. If the startup has high growth curve, rewards could be huge.

Passion: Passion lasts longer than money or title. The internal desire to win at all cost. These people are very hard to find; however can help take your startup reach the promised land of glory. Passionate people make lots of money through their career, more than #1 and #2.

All startups are good.

"Be passionate, rest will follow."

> "Take your passion and make it happen"
>
> from What a Feeling,
> Irene Cara
> in Flashdance

Viral factor measurement

The key ingredients for making a successful viral factor:

1. Email domain.
2. Country.
3. Age.
4. Gender.
5. Language.
6. How many address in the address book?
7. How many invites sent?
8. How many opened?
9. How many clicked?
10. How many converted?
11. If the child converted sent is bigger than the people that signed up, you got a viral factor > 1 which means you are in growth mode.
12. If you don't have a viral factor > 1, then you need to analyze your content and see whether that justifies your existence. You can measure that by how much content per user per time frame is being produced.

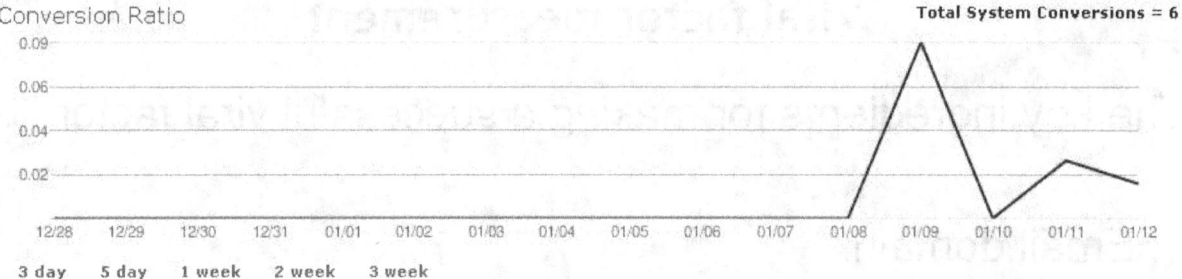

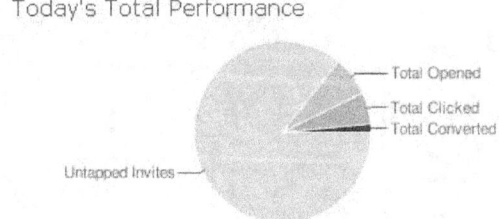

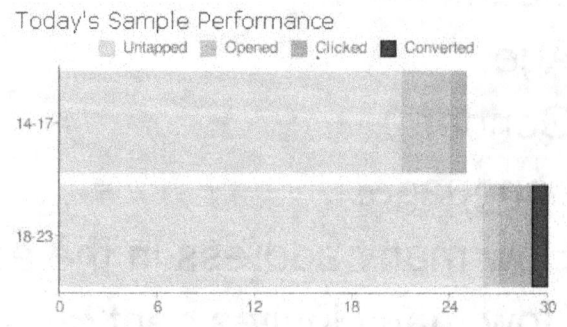

Referrer Tracking for analyzing traffic

Internal referer tracking is important to have real-time information where your traffic is coming from. We have all learned to love Google analytics, but sometimes it's too slow. If you need to know right away the impact of your campaign, you need to have real-time information, captured via dashboard. I am a great believer of metrics. You can find a lot of interesting patterns, and see things way before they will actually happen. Complicated metrics presented in simple terms is an art. Because at the end of the day, it's your investors and founders that have to see them and make important business decisions. The more the metrics, the more visibility into the business. However, most startups don't qualify to have metrics because you need to have traction before applying any real metrics rules.

For mytweetmark, every time the mytweetmark link is clicked, the referrer tracking will tell right away whether google, bing, facebook, twitter, mashable, techcrunch, any location's viralness is accessed and the impact of the campaign. Lots of useful analytics can be built around this fashion. It would also be very interesting from this traffic, mapping CTR to viral factor since every sign up or registration counts into the viral factor. This is definitely something to be investigated and built into analytics in the future.

704	http://googleads.g.doubleclick.net/pagead/ads?client=ca-pub-2861875774931613&output=html&h=280&slotname=5602239005&w=336&lmt=1255967892&flash=10.0.22.87&url=http%3A%2F%2Fwww.activityvillage.co.uk%2Ffree_printables.htm&dt=1256134750062&correlator=1256134750062&frm=0&ga_vid=1026142084.1256134631&ga_sid=1256134631&ga_hid=1056859811&ga_fc=1&u_tz=-300&u_his=6&u_java=1&u_h=600&u_w=800&u_ah=600&u_aw=800&u_cd=32&u_nplug=0&u_nmime=0&biw=771&bih=446&ref=http%3A%2F%2Fwww.google.com%2Fsearch%3Fhl%3Den%26q%3Dfree%2Bbookmark%2Btemplates%2Bfor%2Bkids%26aq%3D1%26oq%3Dfree%2Bbookmark%2Btemplate%26aqi%3DGg3g-m1&fu=0&ifi=1&dtd=47&xpc=gDazrroi4h&p=http%3A//www.activityvillage.co.uk	2009-10-2 09:19:31.1
705	http://www.finance4founders.com/2009/10/19/startup-hiring-i%e2%80%99ve-hired-all-my-friends-now-what/	2009-10-2 09:51:39.6
706	http://www.mytweetmark.com/	2009-10-2 09:51:40.8
707	http://mytweetmark.com/internalReferer/list?offset=680&max=10	2009-10-2 10:30:47.9
708	http://mytweetmark.com/users/list?offset=0&max=10	2009-10-2 10:31:21.5
709	http://mytweetmark.com/myTweetMark/myTweetMarks	2009-10-2 11:07:12.3
710	http://personalweb.about.com/od/searchingandsurfing/a/2bookmarking.htm	2009-10-2 13:46:26.2

Retention measurement

Measuring retention is necessary to figure out user traction. There are a few questions to ask:

1. How many people registered in 30, 60, 90 days?
2. How many came back to the site?
3. Gather retention coefficient by dividing registered with logged in.
4. How many logins per user in those days?
5. How much content?
6. How much social graph?
7. How much communication?

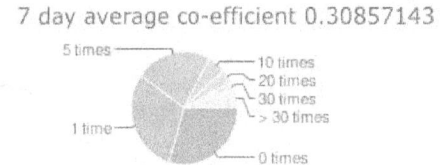

Launching a Product

Product:

Launching a product takes a lot of thought. It takes a long time to get the product market fit. Understanding market and the opening. Building a product and be sound in execution. Investors, customers and users watch the time to market. The design of the product is critical, because it has to match the taste of the audience. Building a brand is crucial. Many battles have to be fought to reach that point. Battles are also validation that something good is happening.

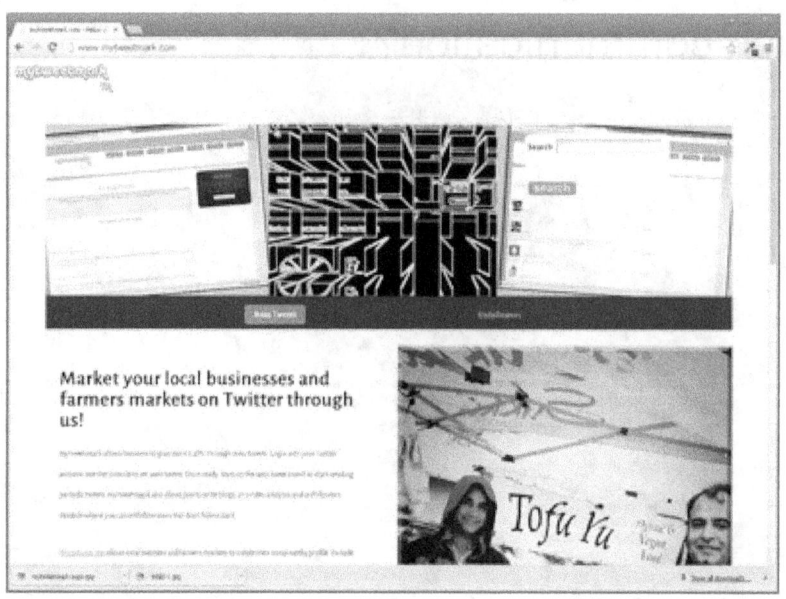

Market:

Recently, when visiting one of the major Angel Investor circuit, where each investor is worth anywhere from $8M - $100M, many entrepreneurs pitched their companies. The funding ask size was anywhere from $500K - $8M for each company. Interestingly enough, to justify where the money would be spent, almost all companies showed in their slide decks that 50% of the money will be used for marketing. Many founders take the 'crank' approach of advertising. As soon as you launch the product, you start burning advertising dollars on major channels, e.g. Facebook ads, Google ads, Twitter ads, and major advertising agency networks. They get really good at AdWords and other tools that allow you to geographically target users by keywords. The problem with this approach is that the traffic generated is artificial. Once the investors get used to seeing traffic numbers increase, they want to continue seeing upticks. Hence the company keeps spending more money. Therefore, these founders looking for funding are justifying 50% of the money generated is for marketing.

There is no right and wrong answer. If the company is like Facebook, it's really easy to generate traffic. It's organic. It's free. Users join, invite their friends. The viral factor, the tick to measure growth in user sign up is high. The retention cohorts, the tick to measure how often the users are coming back to the website is high. The engagement metrics, how much of your features the users are consuming is also high. Hence, investors like to see analytic and data of company growth. They are looking for organic traffic to reduce advertising dollars spent.

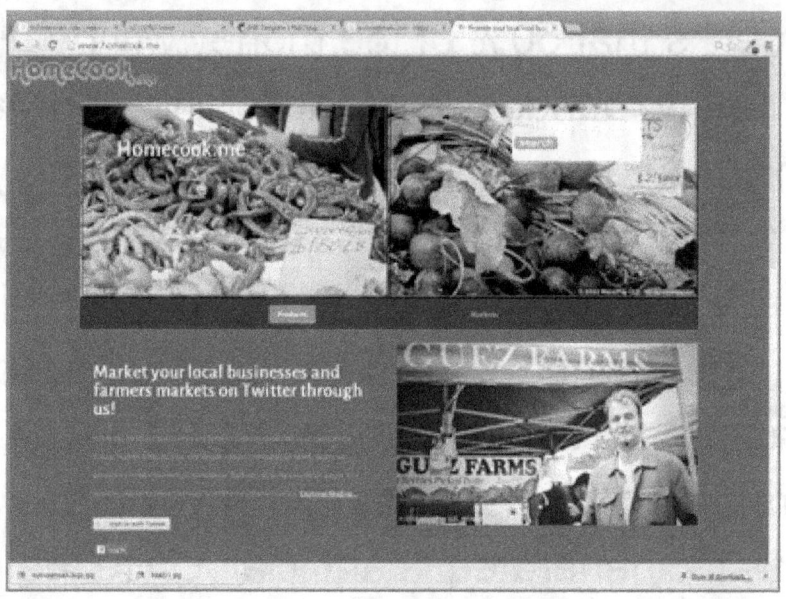

Momentum:

The result of all of this is further analysis of the company as a whole. The product is launched. The marketing and distribution needs to happen to analyze the product. The word needs to be spread. The key networks have to be notified. Email campaigns have to be launched. The internal analytic needs to be implemented to see growth, pageviews, conversions and brand awareness. Positive and negative criticism needs to be evaluated. Some of the negative criticism are validations. Some criticism require further changes in the product for the future. Surrounding the company with best advisers, investors and workforce allows to further evaluate and make critical decisions.

Watch the momentum and then pivot. Change what's not working. Write blogs and figure out free ways for distribution. Reach the most amount of people and educate them on your product. Reach all the free social media channels and create brand awareness.

Building Custom analytics

It is key to obtain key metrics for your company, with information pertaining to business goals and metrics. The tracking can easily be done on emails, buttons and pages. Configure URLs, passing in tracker Ids to each call. Each tracker Id associates with business objects within the system. With each URL call, whether onload of html page, button onclick, or email open, call the tracker URL to increment counters.

Once the tracking data is collected, there are various way to display charts and paginated lists. For each tracker, collect list of tracking, e.g. IP address and information pertaining to clients and browsers. Location information can also be obtained from the IP address. Http headers contain a wealth of user information.

Ab-testing, funnel testing, retention cohorts, engagement metrics and viral factors can be implemented with this mechanism.

Tracker List

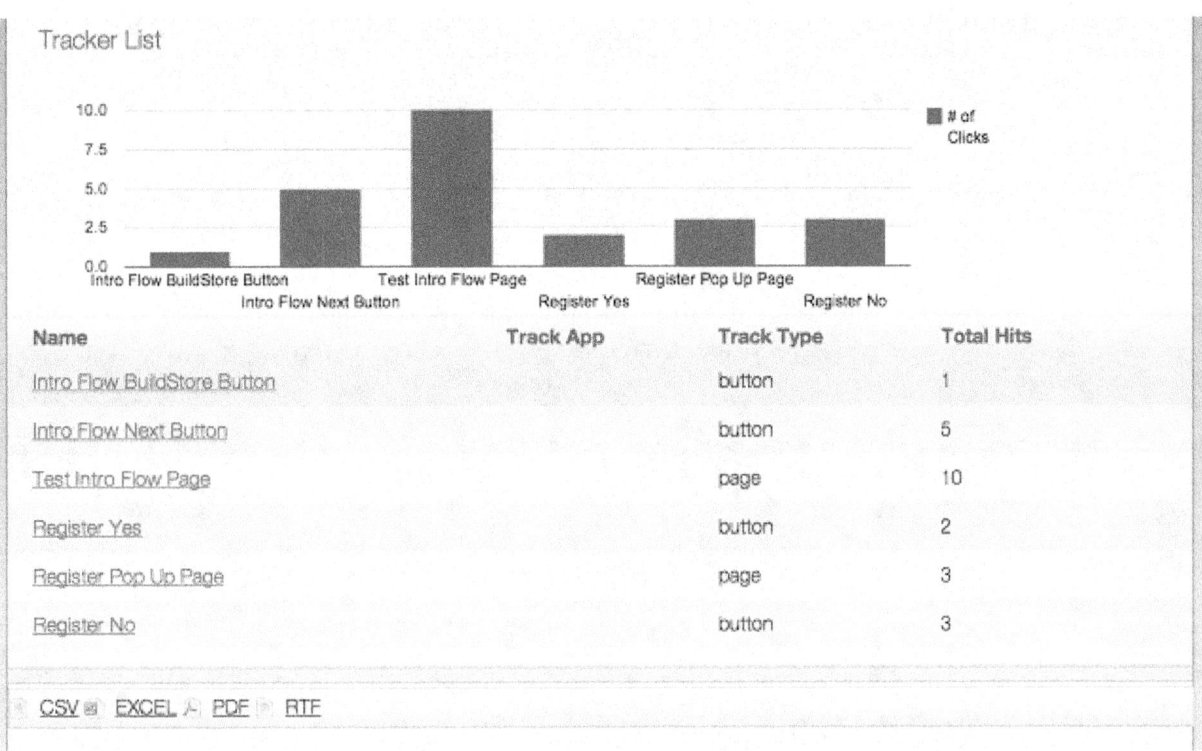

Name	Track App	Track Type	Total Hits
Intro Flow BuildStore Button		button	1
Intro Flow Next Button		button	5
Test Intro Flow Page		page	10
Register Yes		button	2
Register Pop Up Page		page	3
Register No		button	3

CSV   EXCEL   PDF   RTF